Physical chemistry

C000126938

Topic 8 Thermodynamics

Born–Haber cycles

The definitions of enthalpy changes are important and should be learned as they are often asked for in exams. They are also applied in questions on Born–Haber cycles:

- The standard enthalpy of formation ($\Delta_f H^\ominus$) is the enthalpy change when 1 mol of a compound is formed from its elements when all the reactants and products are in their standard states under standard conditions.
- The standard enthalpy of lattice dissociation ($\Delta_L H^\ominus$ or $\Delta_{latt} H^\ominus$) is the enthalpy change when 1 mol of an ionic compound is separated into its component gaseous ions.
- The standard enthalpy of atomisation ($\Delta_a H^\ominus$ or $\Delta_{at} H^\ominus$) is the enthalpy change when 1 mol of gaseous atoms is formed from the element in its standard state.
- The first ionisation enthalpy (or energy) ($\Delta_{IE1} H^\ominus$) is the enthalpy change when 1 mol of electrons is removed from 1 mol of gaseous atoms to form 1 mol of gaseous ions with a single positive charge.
- The second ionisation enthalpy (or energy) ($\Delta_{IE2} H^\ominus$) is the enthalpy change when 1 mol of electrons is removed from 1 mol of gaseous ions with a single positive charge to form 1 mol of gaseous ions with a 2+ charge.
- The third ionisation enthalpy (or energy) ($\Delta_{IE3} H^\ominus$) is the enthalpy change when 1 mol of electrons is removed from 1 mol of gaseous ions with a 2+ charge to form 1 mol of gaseous ions with a 3+ charge.
- The bond dissociation enthalpy is the energy required to break 1 mol of a covalent bond under standard conditions in the gaseous state.
- The first electron affinity is the enthalpy change when 1 mol of gaseous atoms forms 1 mol of gaseous ions with a single negative charge.

- The second electron affinity is the enthalpy change when 1 mol of gaseous ions with a single negative charge forms 1 mol of gaseous ions with a 2– charge.

A Born–Haber cycle links these enthalpy changes and allows the calculation of any change in the cycle. For example, for a group 1 halide:

lattice enthalpy = –enthalpy of formation + enthalpy of atomisation of metal + first ionisation enthalpy + enthalpy of atomisation of non-metal + first electron affinity

For example, for a group 2 halide:

lattice enthalpy = –enthalpy of formation + enthalpy of atomisation of metal + first ionisation energy + second ionisation enthalpy + 2 × enthalpy of atomisation of non-metal + 2 × first electron affinity

The bond dissociation enthalpy or the bond enthalpy is equal to 2 × enthalpy of atomisation of the non-metal.

The experimental value of lattice enthalpy is determined from the Born–Haber cycle. The perfect ionic model (assuming point charges that are perfect spheres) gives a theoretical value for the lattice enthalpy. If the experimental and theoretical values for lattice enthalpy are very close together, the bonding in the compound is ionic. If the experimental value is greater than the theoretical value, then there is additional covalent bonding in the compound.

The enthalpy of solution is the enthalpy change when 1 mol of a solute dissolves in water. The hydration enthalpy is the enthalpy change when 1 mol of gaseous ions is converted into 1 mol of aqueous ions. The enthalpy of solution = lattice dissociation enthalpy + total of enthalpies of hydration.

---

1 **Which one of the following changes is exothermic? (AO1)**   **1 mark**

   A $Na(s) \rightarrow Na(g)$

   B $Na(g) \rightarrow Na^+(g) + e^-$

   C $O(g) + e^- \rightarrow O^-(g)$

   D $O^-(g) + e^- \rightarrow O^{2-}(g)$

2 The Born–Haber cycle for potassium chloride is shown below.

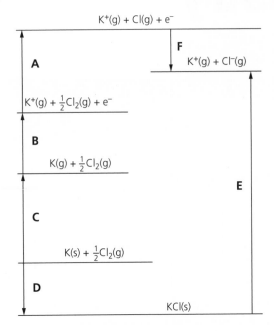

a   Which letter represents the enthalpy of atomisation of potassium? (AO3) `1 mark`

....

....

b   Which letter represents the enthalpy of lattice dissociation of potassium chloride? (AO3) `1 mark`

....

....

c   Which letter represents the first ionisation energy of potassium? (AO3) `1 mark`

....

....

d   Which letter represents the first electron affinity of chlorine? (AO3) `1 mark`

....

....

e   The following values are given:

Enthalpy change	$\Delta H/\text{kJ mol}^{-1}$
A	+121
B	+420
C	+89
D	−437
E	+703

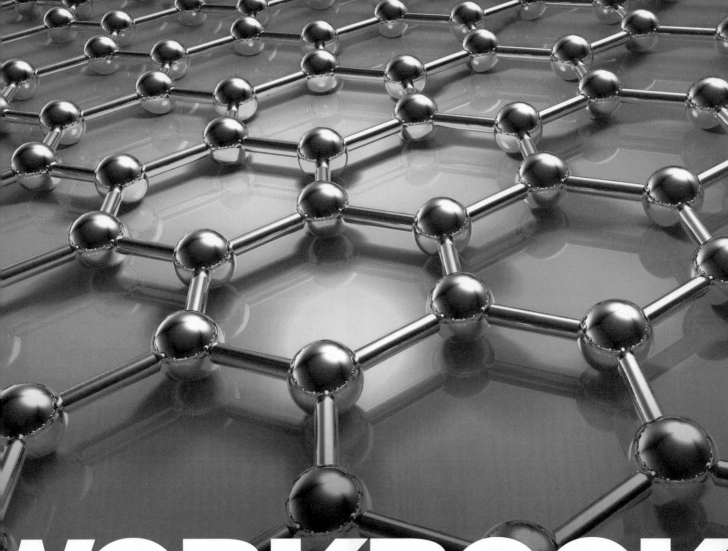

WORKBOOK

Document ③

Chemistry

Physical chemistry 2

Alyn G. McFarland and Nora Henry

FOR THE
2015
SPECIFICATIONS

HODDER
EDUCATION
LEARN MORE

Contents

Physical chemistry

(1) **This workbook will help you** to prepare for the following exams:

- AQA Chemistry A-level Paper 1: the exam is 2 hours long, worth 105 marks and 35% of your A-level. The exam is made up of short- and long-answer questions.
- AQA Chemistry A-level Paper 2: the exam is 2 hours long, worth 105 marks and 35% of your A-level. The exam is made up of short- and long-answer questions.
- AQA Chemistry A-level Paper 3: the exam is 2 hours long, worth 90 marks and 30% of your A-level. The exam includes questions testing across the whole specification.

(2) **For each topic** of each section there are:

- stimulus materials including key terms and concepts
- short-answer questions
- long-answer questions
- questions that test your mathematical skills
- space for you to write

(3) **Answering the questions** will help you to build your skills and meet the assessment objectives (AO1) (knowledge and understanding), AO2 (application) and AO3 (analysis, interpretation and evaluation).

(4) **You still need to** read your textbook and refer to your revision guides and lesson notes.

(5) **Marks** available are indicated for all questions so that you can gauge the level of detail required in your answers.

(6) **Timings** are given for the exam-style questions to make your practice as realistic as possible.

(7) **Answers** are available at: www.hoddereducation.co.uk/workbookanswers

Calculate a value for F. (AO2)

3 marks

....

....

....

....

....

③ The table below gives data for the Born–Haber cycle of magnesium oxide.

Enthalpy change	ΔH/kJ mol^{-1}
Enthalpy of atomisation of magnesium	+150
First ionisation energy of magnesium	+736
Second ionisation energy of magnesium	+1450
Bond dissociation enthalpy of oxygen	+496
First electron affinity of oxygen	−142
Second electron affinity of oxygen	+844
Enthalpy of formation of magnesium oxide	−602

a Write equations, including state symbols, which represent the following changes in the Born–Haber cycle. (AO3)

   i first ionisation energy of magnesium

1 mark

   ....

   ii second ionisation energy of magnesium

1 mark

   ....

   iii enthalpy of formation of magnesium oxide

1 mark

   ....

   iv second electron affinity of oxygen

1 mark

   ....

   v enthalpy of lattice dissociation of magnesium oxide

1 mark

   ....

5

b The Born–Haber cycle below is for magnesium oxide.

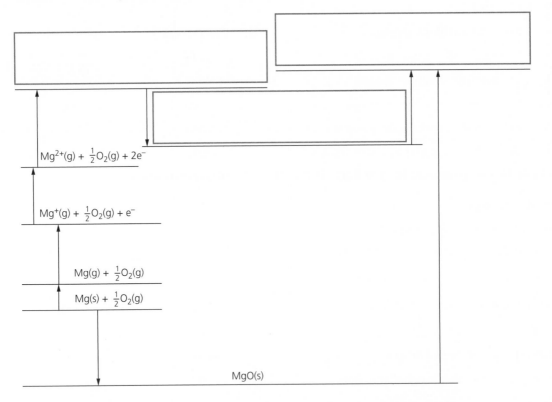

i Complete the Born–Haber cycle by showing what is present at the three empty levels. (AO3)

3 marks

ii Using the constructed Born–Haber cycle, or any other method, calculate the enthalpy of lattice dissociation of magnesium oxide. (AO2)

3 marks

4 The table and diagram below give the enthalpy changes and the Born–Haber cycle for aluminium oxide.

	Name of enthalpy change	$\Delta H^{\ominus}$/kJ mol^{-1}
A	Enthalpy of atomisation of oxygen	+248
B	First electron affinity of oxygen	−141
C	Second electron affinity of oxygen	+790
D	First ionisation enthalpy of aluminium	+580
E	Second ionisation enthalpy of aluminium	+1800
F	Third ionisation enthalpy of aluminium	*To be calculated*
G	Enthalpy of atomisation of aluminium	+324
H	Enthalpy of lattice dissociation of aluminium oxide	+15180
I	Enthalpy of formation of aluminium oxide	−1676

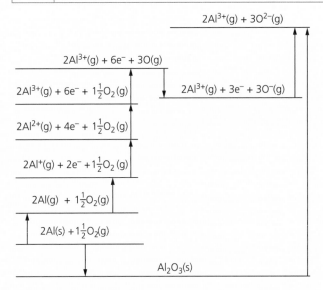

a Label the enthalpy changes on the Born–Haber cycle using the letters A–I. (AO1)   4 marks

b Calculate a value for the third ionisation enthalpy of aluminium. (AO2)   4 marks

....
....
....
....
....
....
....
....
....
....
....
....

5 The diagram below shows the Born–Haber cycle for magnesium chloride.

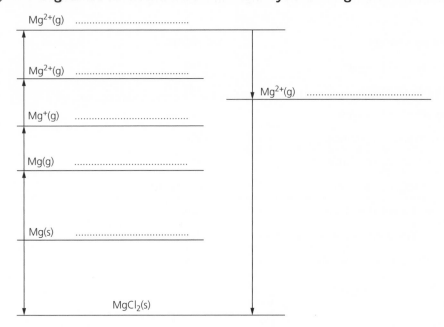

Mg²⁺(g) ....

Mg²⁺(g) ....

Mg²⁺(g) ....

Mg⁺(g) ....

Mg(g) ....

Mg(s) ....

MgCl₂(s)

a Complete this Born–Haber cycle for magnesium chloride, giving the species present at each level. Include state symbols. (AO2)　　6 marks

b The table below gives some enthalpy values associated with magnesium chloride.

	Enthalpy change/kJ mol⁻¹
Enthalpy of atomisation of magnesium	+150
First electron affinity of chlorine	−364
First ionisation energy of magnesium	+736
Second ionisation energy of magnesium	+1450
Enthalpy of formation of magnesium chloride	−642
Lattice enthalpy of formation of magnesium chloride	−2492
Enthalpy of hydration of magnesium ions	−1920
Enthalpy of hydration of chloride ions	−364

i Use the data in the table above to calculate a value for the enthalpy of atomisation of chlorine. (AO2)　　3 marks

....

....

....

....

....

ii Calculate the enthalpy of solution of magnesium chloride. (AO2)　　3 marks

....

....

....

....

....

iii State and explain whether you would expect a calcium ion to have a higher or lower hydration enthalpy than a magnesium ion. (AO1) `2 marks`

....

....

....

....

Exam-style question

(15)

6 **a** The enthalpy changes shown below relate to magnesium fluoride.

Enthalpy change	$\Delta H/\text{kJ mol}^{-1}$
$Mg(s) \rightarrow Mg(g)$	+150
$Mg(g) \rightarrow Mg^+(g) + e^-$	+736
$Mg^+(g) \rightarrow Mg^{2+}(g) + e^-$	+1450
$F_2(g) \rightarrow 2F(g)$	+158
$F(g) + e^- \rightarrow F^-(g)$	−348
$MgF_2(s) \rightarrow Mg^{2+}(g) + 2F^-(g)$	+2883
$MgF_2(s) \rightarrow Mg^{2+}(aq) + 2F^-(aq)$	−20
$Mg^{2+}(g) \rightarrow Mg^{2+}(aq)$	−1891

i Calculate the enthalpy of formation of magnesium fluoride. `3 marks`

....

....

....

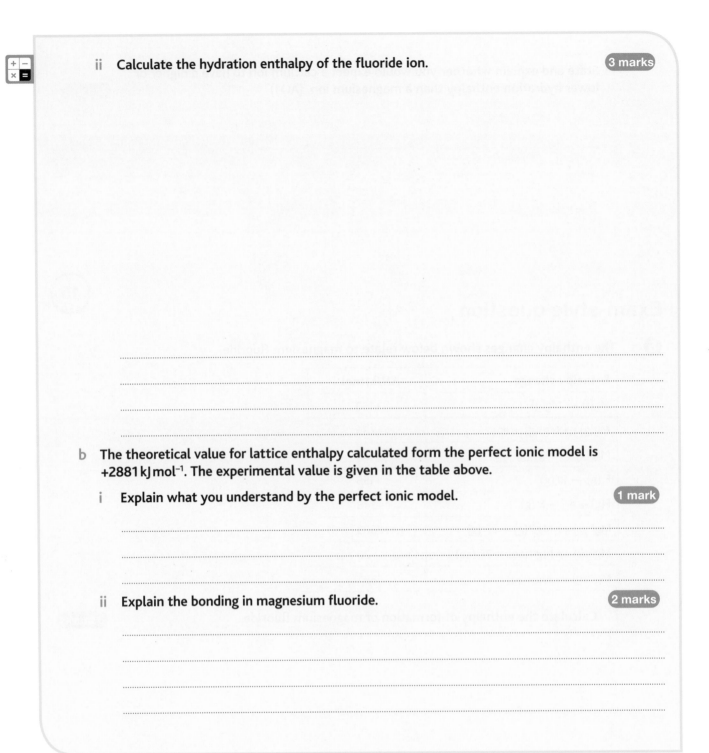

ii   Calculate the hydration enthalpy of the fluoride ion.   **3 marks**

b   The theoretical value for lattice enthalpy calculated form the perfect ionic model is +2881 kJ mol⁻¹. The experimental value is given in the table above.

i   Explain what you understand by the perfect ionic model.   **1 mark**

ii   Explain the bonding in magnesium fluoride.   **2 marks**

Gibbs free-energy change (ΔG) and entropy change (ΔS)

Entropy is a measure of disorder and is represented by S°. ΔS° is the change in entropy. The units of entropy and ΔS° are $J\,K^{-1}\,mol^{-1}$.

$$\Delta S^\circ = \Sigma S^\circ \text{ (products)} - \Sigma S^\circ \text{ (reactants)}$$

Gases have a higher entropy value than liquids/ solutions, which have a higher entropy value than solids. Highly ordered solids such as ionic compounds, macromolecular substances and metals have very low entropy values. A reaction that produces a gas would generally show an increase in entropy.

$\Delta G^\circ = \Delta H^\circ - T\Delta S^\circ$, where ΔG° is Gibbs free energy $(kJ\,mol^{-1})$, ΔH° is the enthalpy change in the reaction $(kJ\,mol^{-1})$, T is the temperature in kelvin (K), ΔS° is the change in entropy in the reaction $(kJ\,K^{-1}\,mol^{-1})$. In the ΔG° expression ΔS° has to be divided by 1000 to convert it from $J\,K^{-1}\,mol^{-1}$ to $kJ\,K^{-1}\,mol^{-1}$. For a reaction to be feasible ΔG° must be equal to zero or less than zero.

A reaction that is exothermic and shows an increase in entropy is feasible at any temperature. A reaction that is endothermic and shows a decrease in entropy is *not* feasible at any temperature. A reaction that is exothermic and shows a decrease in entropy is feasible below a certain temperature. A reaction that is endothermic and shows an increase in entropy is feasible above a certain temperature.

A graph of ΔG against T shows a gradient that is $-\Delta S$ and the intercept with the ΔG axis is ΔH.

At a change of state $\Delta G^{\ominus} = 0$. This can be used to determine the melting or boiling point of a substance or the enthalpy change of fusion (change from solid to liquid) or vaporisation (change from liquid to gas).

1 Which of the following reactions shows a decrease in entropy? (AO3)   1 mark

   A $2Na(s) + 2H_2O(l) \rightarrow 2NaOH(aq) + H_2(g)$

   B $2Na(s) + Cl_2(g) \rightarrow 2NaCl(s)$

   C $Na_2CO_3(s) + 2HCl(aq) \rightarrow 2NaCl(aq) + CO_2(g) + H_2O(l)$

   D $NaCl(s) + H_2SO_4(l) \rightarrow NaHSO_4(s) + HCl(g)$

2 Which of the following does not have units of $kJ\,mol^{-1}$? (AO1)   1 mark

   A $\Delta_f H^{\ominus}$

   B $\Delta S^{\ominus}$

   C $\Delta G^{\ominus}$

   D $\Delta_L H^{\ominus}$

3 Which one of the following reactions would be feasible at any temperature? (AO1)   1 mark

   A an endothermic reaction with a decrease in entropy

   B an endothermic reaction with an increase in entropy

   C an exothermic reaction with a decrease in entropy

   D an exothermic reaction with an increase in entropy

4 When lead(II) nitrate is heated it decomposes according to the equation:

   $2Pb(NO_3)_2(s) \rightarrow 2PbO(s) + 4NO_2(g) + O_2(g)$

The table below gives the absolute entropy values and the enthalpy of formation for all the substances in the reaction.

Substance	$Pb(NO_3)_2(s)$	$PbO(s)$	$NO_2(g)$	$O_2(g)$
$S^{\ominus}/J\,K^{-1}\,mol^{-1}$	217.9	68.7	240	205
$\Delta_f H^{\ominus}/kJ\,mol^{-1}$	−451.9	−217.3	+33.2	0

   a Explain why the enthalpy of formation of oxygen is zero. (AO1)   1 mark b   Calculate $\Delta S^{\ominus}$ for this reaction. (AO2)   3 marks

....

....

....

....

c   Calculate $\Delta H^{\ominus}$ for this reaction. (AO2)   3 marks

....

....

....

....

d   Show that this reaction is not feasible at 500 K. (AO2)   4 marks

....

....

....

....

e   Calculate the temperature at which this reaction becomes feasible. (AO2)   3 marks

....

....

....

5 The enthalpy of fusion (melting) of magnesium is +9 kJ mol^{-1}. The absolute entropy values of magnesium as a solid and a liquid are 32.7 and 42.7 J K^{-1} mol^{-1}. Calculate the melting point of magnesium in kelvin (K). (AO2)   4 marks

....

....

....

....

6 The table below shows the absolute entropy values and enthalpies of formation of some substances in different states.

Substance	$S^{\ominus}$/J K^{-1} mol^{-1}	$\Delta_f H^{\ominus}$/kJ mol^{-1}
Al(s)	28	0
Al(l)	40	+11
H$_2$O(s)	41	−292
H$_2$O(l)	70	−286
H$_2$O(g)	189	−242
CO$_2$(g)	214	−394
Al$_2$O$_3$(s)	51	−1676
H$_2$(g)	131	0
O$_2$(g)	205	0
CO(g)	198	−111
Mg(s)	34	0
MgO(s)	27	−601
MgO(l)	48	To be calculated

a   Aluminium reacts with water vapour according to the equation:

$$2Al(s) + 3H_2O(g) \rightarrow Al_2O_3(s) + 3H_2(g)$$

   i   Calculate the enthalpy change for this reaction. (AO2)   (3 marks)

   ii Calculate the entropy change for this reaction. (AO2)   (3 marks)

   iii Show that the reaction of aluminium with water vapour is feasible at 500 K.   (3 marks)

b   The melting point of magnesium oxide is 3125 K. Calculate a value for the enthalpy of formation of MgO(l). (AO2)   (6 marks) c   Water vapour can decompose into hydrogen and oxygen:

$$H_2O(g) \rightarrow H_2(g) + \tfrac{1}{2}O_2(g)$$

Calculate the temperature at which this decomposition occurs. (AO2)   4 marks

....

....

....

....

....

....

....

Exam-style question

⑮

7 When heated strongly, copper(II) sulfate decomposes according to the equation:

$$CuSO_4(s) \rightarrow CuO(s) + SO_3(g)$$

a   Plot the following values of $\Delta G^{\ominus}$ against temperature, T.   `3 marks`

T/K	400	800	1200	1600
$\Delta G^{\ominus}/kJ\,mol^{-1}$	+145	+71	−4	−78

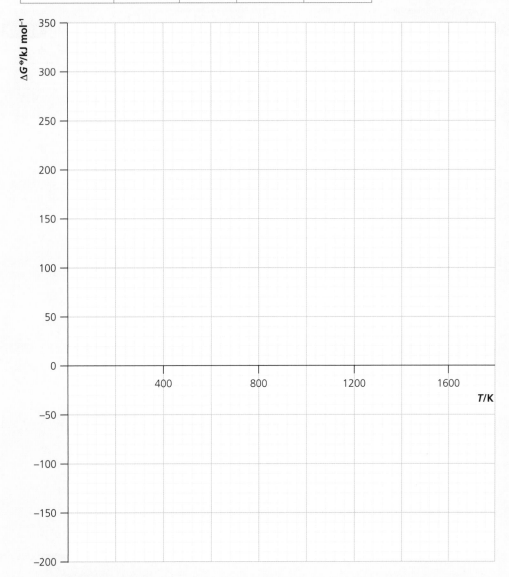

b   Calculate the gradient of the line you have drawn. Give your answer to
3 significant figures.   `2 marks`

....

....

....

c   What is the intercept on the $\Delta G^{\ominus}$ axis?   `1 mark`

....

....

d State a value for $\Delta H^{\ominus}$ for this reaction. Include units in your answer.

2 marks

....

....

e State a value for $\Delta S^{\ominus}$ for this reaction. Include units in your answer.

2 marks

....

....

....

f At what temperature does the reaction become feasible?

1 mark

....

....

Topic 9 Rate equations

A rate equation links the rate of reaction with the rate constant and the concentration of the reactants. For the reaction:

$$aA + bB \rightarrow cC + dD$$

the rate equation is:

$$\text{rate} = k[A]^x[B]^y$$

where k is the rate constant, [A] is the concentration of A, [B] is the concentration of B, x is the order with respect to A and y is the order with respect to B. The overall order of reaction is $x + y$. Orders are 0, 1 or 2.

A zero-order reactant does take not part in the rate-determining step in the mechanism for the reaction. A zero-order reactant may be left out of the rate equation. Orders may be determined experimentally using rate and concentration data.

The units of the rate constant depend on the overall order of the reaction.

For a reaction in which there is a directly measurable reactant (such as a coloured substance), one experiment may be carried out to determine the order with respect to this reactant. A graph of concentration against time gives one of three shapes:

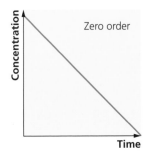

Where there is a directly measurable product, gas volume, mass measurements, colorimetry or sample quench titration methods may all be used to determine the rate of reaction. Where the reactant is not directly measurable, a series of experiments will be set up changing the concentration of one reactant and measuring initial rate. A graph of initial rate against concentration gives one of three shapes:

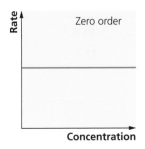

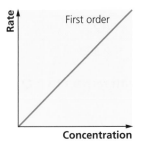

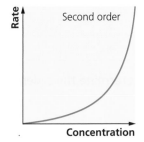

For a zero order reactant the graph of rate will be a straight line where concentration has no effect on the rate of reaction.

For a first order reactant the graph of rate will be a straight line where as the concentration of the reactant doubles the rate of reaction also doubles.

For a second order reactant the graph of rate will be a curved line where as the concentration of the reactant doubles the rate of reaction quadruples (×4).

The Arrhenius equation:

$$k = Ae^{-\frac{E_a}{RT}}$$

links the rate constant (k), temperature (T) and activation energy (E_a). A is the Arrhenius constant, R is the gas constant ($8.31\,J\,K^{-1}\,mol^{-1}$) and e is a universal constant (2.71828). A graph of $\ln k$ against $1/T$ gives a straight line with an intercept on the $\ln k$ axis of $\ln A$. A may be determined by $e^{(\ln A)}$. A graph of $\ln k$ against $1/T$ has a gradient of $-E_a/R$. E_a determined from this gradient from this type of graph will have units of $J\,mol^{-1}$.

1 A chemical reaction is given as:

$$2A + B \rightarrow 3C$$

The rate equation for the reaction is rate = $k[A][B]$.

Which of the following are the units of the rate constant? (AO1)

A s^{-1}

B $mol^{-1} dm^3 s^{-1}$

C $mol^{-2} dm^6 s^{-1}$

D $mol^{-3} dm^9 s^{-1}$

2 The equation for the reaction $X + 2Y \rightarrow Z$ is:

$$rate = k[Y]^2$$

Which one of the following would show a correct mechanism based on this information? (AO3)   1 mark

A $X + Y \rightarrow A$ (slow step) $A + Y \rightarrow Z$ (fast step)

B $X \rightarrow B + C$ (slow step) $B + C + 2Y \rightarrow Z$ (fast step)

C $2X + 2Y \rightarrow D$ (slow step) $D \rightarrow Z + X$ (fast step)

D $2Y \rightarrow E$ (slow step) $E + X \rightarrow Z$ (fast step)

3 For the chemical reaction $2P + 3Q + R \rightarrow 2S$ a series of experiments were carried out. The results are shown in the table below.

Experiment	[P]/mol dm^{-3}	[Q]/mol dm^{-3}	[R]/mol dm^{-3}	Rate/mol dm^{-3} s^{-1}
1	0.154	0.154	0.154	1.24×10^{-4}
2	0.308	0.154	0.154	2.48×10^{-4}
3	0.231	0.231	0.154	1.86×10^{-4}
4	0.231	0.308	0.308	7.44×10^{-4}

a Determine the order of reaction with respect to P, Q and R. (AO2)   3 marks

b Write a rate equation for this reaction. (AO3)   1 mark

c Use the results of experiment 1 to calculate a value for the rate constant, k.
Give your answer to 3 significant figures and state its units. (AO2) `3 marks`

4 The Arrhenius equation is shown below.

$$k = Ae^{-\frac{E_a}{RT}}$$

a State the meaning of each of the terms in the expression. (AO1) `6 marks`

k

A

e

E_a

R

T

b Describe how the value of k changes as T increases. (AO1) `1 mark`

c A graph of $\ln k$ against $1/T$ was drawn. The gradient was determined to be −24 500 and the intercept of the line with the vertical axis was 44.0.

i Write the equation for the straight line obtained in terms of $\ln k$ and $1/T$. (AO1) `1 mark`

ii Calculate a value for A and calculate a value for E_a in kJ mol^{-1}. ($R = 8.31$ J K^{-1} mol^{-1}).
Give both answers to 3 significant figures. (AO2) `4 marks`

⑤ The graph below shows $\ln k$ plotted against $1/T$ for the reaction A + 2B → 3C + D.

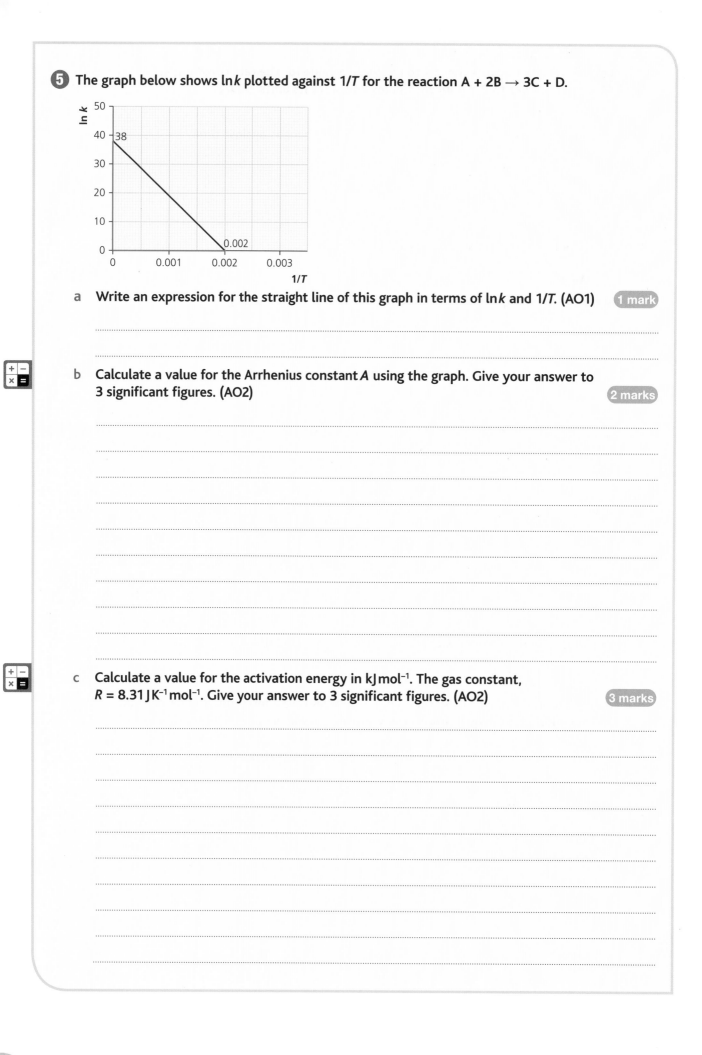

a Write an expression for the straight line of this graph in terms of $\ln k$ and $1/T$. (AO1)　1 mark

b Calculate a value for the Arrhenius constant A using the graph. Give your answer to 3 significant figures. (AO2)　2 marks

c Calculate a value for the activation energy in $kJ\,mol^{-1}$. The gas constant, $R = 8.31\,J\,K^{-1}\,mol^{-1}$. Give your answer to 3 significant figures. (AO2)　3 marks

6 A undergoes a decomposition reaction:

   A → B + C

The graph below shows how the concentration of A changes against time.

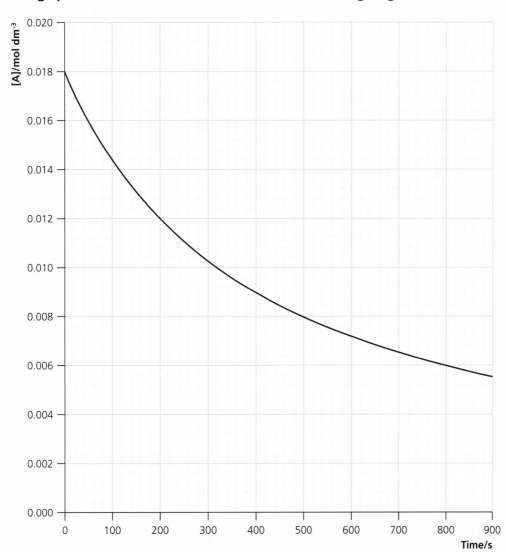

a i State the time at which the concentration of A had decreased by half. (AO3)

....

....

....

....

ii Explain why the order of reaction with respect to A cannot be zero. (AO1) 1 mark

....

....

....

....

b Two tangents are drawn on the graph as shown below. Tangent 1 is drawn at 0.018 mol dm⁻³ and tangent 2 at 0.012 mol dm⁻³.

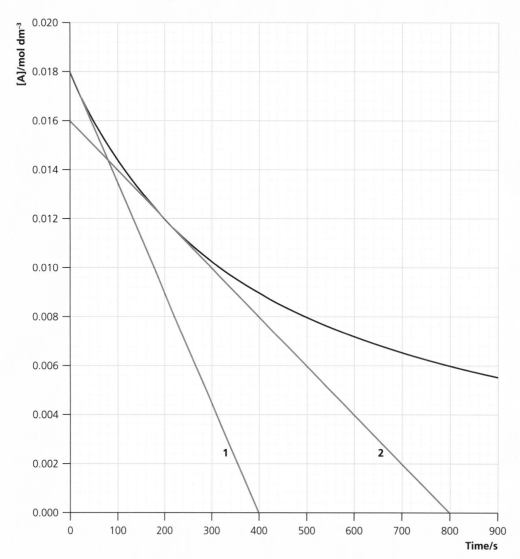

i Using the graph determine the rate of reaction when the concentration of A is 0.012 mol dm⁻³ and 0.018 mol dm⁻³. (AO3)

4 marks

....

....

....

....

....

....

....

....

....

....

ii Determine the order of reaction with respect to A. Write a rate equation and use the data determined at 0.018 mol dm⁻³ to calculate a value for the rate constant to 3 significant figures. State the units of the rate constant. (AO2)   6 marks

....
....
....
....
....
....
....
....
....
....
....
....

Exam-style question

20

7 For the chemical reaction:

$A + B + 2C \rightarrow D + 3E$

the rate equation is rate = $k[B][C]^2$

The table below gives data from a series of experiment carried out.

Experiment	[A]/mol dm⁻³	[B]/mol dm⁻³	[C]/mol dm⁻³	Rate/mol dm⁻³ s⁻¹
1	0.0144	0.0230	0.0128	2.20×10^{-3}
2	0.0180	0.0345	0.0256	*To be calculated*
3	0.0180	*To be calculated*	0.0192	9.90×10^{-3}
4	0.0288	0.0184	*To be calculated*	1.10×10^{-2}

a   Calculate the value for the rate in experiment 2.   4 marks

....
....
....
....
....
....
....

b Calculate a value for the concentration of B in experiment 3. (4 marks)

....

....

....

....

....

....

....

c Calculate a value for the concentration of C in experiment 4. (4 marks)

....

....

....

....

....

....

....

d Using experiment 1 determine a value for the rate constant, *k*. Give your answer to 3 significant figures. State its units. (3 marks)

....

....

....

....

....

....

....

Topic 10 Equilibrium constant (K_p) for homogeneous systems

For the reaction $aB(g) + bB(g) \rightleftharpoons cC(g) + dD(g)$:

$$K_p = \frac{(pC)^c(pD)^d}{(pA)^a(pB)^b}$$

where pC represents the partial pressure of C etc. In a gaseous homogeneous equilibrium, the mole fraction of a gas is calculated as the number of moles of that gas divided by the total number of moles of gas at equilibrium. For example, the mole fraction of C:

$$x_C = \frac{\text{moles of C present at equilibrium}}{\text{total moles of gas present at equilibrium}}$$

The sum of all the mole fractions for all gases present at equilibrium should be 1. The partial pressures are calculated by multiplying the mole fraction by the total pressure. The total pressure is often represented by P, so $pC = x_C \times P$ (partial pressure = mole fraction × total pressure).

K_p may have no units if the pressure units cancel out on the top and bottom of the expression, but if it does have units they will be based on the units of the total pressure (usually kPa or Pa).

As it is an equilibrium constant, the only factor that affects the value of K_p is a change in temperature.

Calculations may determine a value for K_p, or a value for K_p may be given and amounts at equilibrium or a value for the overall pressure (P) determined.

---

1 An equilibrium mixture was found to contain 0.0120 mol of nitrogen, 0.0740 mol of hydrogen and 0.114 mol of ammonia. Which of the following is the partial pressure of nitrogen in the mixture if the total pressure is 100 kPa? (AO2) **1 mark**

A 1.2

B 6

C 37

D 57

....

2 a Write an expression for K_p for the following homogeneous gaseous reactions at equilibrium. (AO1)

  i $N_2(g) + 3H_2(g) \rightleftharpoons 2NH_3(g)$ **1 mark**

....

  ii $2HI(g) \rightleftharpoons H_2(g) + I_2(g)$ **1 mark**

....

iii $4NH_3(g) + 5O_2(g) \rightleftharpoons 4NO(g) + 6H_2O(g)$

1 mark

iv $SO_2(g) + \frac{1}{2}O_2(g) \rightleftharpoons SO_3(g)$

1 mark

b For which one of the reactions in 2a would K_p have no units? (AO3)

1 mark

c Explain how an increase in pressure would affect the position of equilibrium and the value of K_p in reaction i. (AO1/AO3)

4 marks

d The total pressure in reaction iv was in Pa. What are the units of K_p? (AO3)

1 mark

3 A mixture of 0.0240 mol of sulfur dioxide and 0.0200 mol of oxygen was allowed to reach equilibrium at a pressure of 55 kPa.

$2SO_2(g) + O_2(g) \rightleftharpoons 2SO_3(g)$

At equilibrium 15.0% of the sulfur dioxide had reacted. Calculate the moles of each gas present at equilibrium. (AO2)

3 marks

b Calculate the mole fraction of SO_2, O_2 and SO_3 in this equilibrium mixture and from here calculate the partial pressures. (AO2)   6 marks

....

....

....

....

....

....

....

....

....

c Write an expression for K_p for this reaction and calculate its value to 3 significant figures. State the units. (AO2)   4 marks

....

....

....

....

....

....

....

....

4 In the equilibrium below:

$$2A(g) \rightleftharpoons B(g) + C(g)$$

1.00 mol of A was allowed to reach equilibrium at 350 K. K_p for this reaction at 350 K is 0.0400. Calculate the amount, in moles, of A, B and C present in the equilibrium mixture. (AO2)   6 marks

....

....

....

....

....

....

....

....

....

....

27

5 For the equilibrium $2P(g) + Q(g) \rightleftharpoons 3R(g)$, 0.247 mol of P and 0.173 mol of Q were mixed at 500 K and allowed to reach equilibrium. The equilibrium mixture was found to contain 0.0930 mol of R. The total pressure was 105 kPa. Determine a value for K_p at 500 K and state any units of K_p. Give your answer to 3 significant figures. (AO2)   `6 marks`

Topic 11 Electrode potentials and electrochemical cells

Standard electrode potentials give a measure of the feasibility of a half-cell reaction. The value given is in volts (V).

A half-cell may consist of:

- a metal dipping into a solution of its ions: for example, Cu dipping into a solution of Cu^{2+} ions
- a gas in contact with a solution of its ions using a platinum electrode: for example, $Cl_2(g)$ bubbled into a solution containing chloride ions with a platinum electrode in the solution
- two ions both in solution with a platinum electrode to allow contact: for example, Fe^{2+} and Fe^{3+} ions both in the same solution with platinum metal dipping into the solution

When two half-cells are combined, the overall cell has one half-cell in which oxidation occurs (drawn as the left-hand half-cell) and one half-cell in which reduction occurs (drawn as the right-hand half-cell). The overall voltage of a cell is called the electromotive force (emf).

A positive emf indicates a feasible reaction, whereas a negative emf indicates a reaction that is not feasible.

The standard conditions used for all cells is 298 K, 100 kPa pressure and any solutions at 1.00 mol dm^{-3}.

The standard hydrogen electrode is used to measure standard electrode potentials. It consists of $H_2(g)$ at 100 kPa pressure, a platinum electrode and a solution of $H^+(aq)$ of concentration 1.00 mol dm^{-3}. It should be operated at 298 K.

The strongest reducing agent (reductant) is the one, from a list, that is most easily oxidised. The strongest oxidising agent (oxidant) is the one that is most easily reduced.

The negative electrode in a cell is the one at which the oxidation reaction occurs.

The conventional (IUPAC) cell representation shows the oxidation half-cell on the left and the reduction half-cell on the right separated by a salt bridge indicated by ‖. Substances in a half-cell in the same state or phase are separated by a comma; substances in a different state or phase are separated by a phase boundary line, |.

Commercial electrochemical cells are either primary cells (which are non-rechargeable) or secondary cells (which are rechargeable) or fuel cells. Fuels cells maintain a steady voltage as long as they are supplied with a fuel such as hydrogen or ethanol.

---

1 From the following standard electrode potentials:

Electrode half-equation	$E^\ominus$/V
$Mg^{2+}(aq) + 2e^- \rightarrow Mg(s)$	−2.37
$Fe^{2+}(aq) + 2e^- \rightarrow Fe(s)$	−0.44
$V^{2+}(aq) + 2e^- \rightarrow V(s)$	−0.26
$Ni^{2+}(aq) + 2e^- \rightarrow Ni(s)$	−0.25
$Cu^{2+}(aq) + 2e^- \rightarrow Cu(s)$	+0.34

Which is the strongest reducing agent? (AO3)

1 mark

A copper

B copper(ii) ions

C magnesium

D magnesium ions

2 The following diagram shows a typical primary cell.

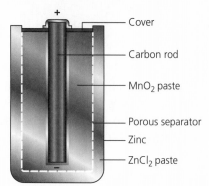

Cover
Carbon rod
MnO_2 paste
Porous separator
Zinc
$ZnCl_2$ paste

a Write an equation for the reaction occurring at the negative electrode. (AO2) 1 mark

b The reaction occurring at the positive electrode is:

$$MnO_2 + H_2O + e^- \rightarrow MnO(OH) + OH^-$$

Explain, using oxidation states, why this is a reduction reaction. (AO3) 3 marks

c Write an overall equation for the reaction occurring in this cell. (AO2) 2 marks

d Suggest why, after prolonged use, the cell may leak. (AO1) 1 mark

3 The table below shows some standard electrode potentials.

Standard half-cell equation	$E^{\ominus}$/V
$Zn^{2+}(aq) + 2e^- \rightarrow Zn(s)$	−0.76
$Fe^{2+}(aq) + 2e^- \rightarrow Fe(s)$	−0.44
$Sn^{2+}(aq) + 2e^- \rightarrow Sn(s)$	−0.14
$Fe^{3+}(aq) + e^- \rightarrow Fe^{2+}(aq)$	+0.77
$Cr_2O_7^{2-}(aq) + 14H^+(aq) + 6e^- \rightarrow 2Cr^{3+}(aq) + 7H_2O(l)$	+1.33

a The first step in the formation of rust is the oxidation of iron to Fe^{2+} by atmospheric oxygen. Explain how the presence of zinc prevents rusting. Calculate the emf of any cell considered. (AO3) `3 marks`

....

....

....

....

....

....

....

b A tin food can is an iron can coated in tin. The surface of the tin is coated with tin(II) oxide, which contains Sn^{2+} ions. Explain why the iron metal under the tin would oxidise to Fe^{2+} if the tin coating was damaged or scratched, exposing the iron. Calculate the emf of any cell considered. (AO3) `3 marks`

....

....

....

....

....

....

c Explain how the potential of the Fe^{3+}/Fe^{2+} half-cell could be measured. (AO1) `5 marks`

....

....

....

....

....

....

....

....

....

d Tin will reduce iron(III) ions to iron(II) ions but will not reduce iron(II) ions to iron.

    i Explain why tin will reduce iron(III) ions to iron(II) but will not reduce iron(II) ions to iron. Calculate the emf of any cells considered. (AO3) `4 marks`

    ii Give the conventional cell notation for the cell formed between Sn^{2+}/Sn and Fe^{3+}/Fe^{2+} half-cells. (AO2) `2 marks`

    iii Write an ionic equation for the reaction between tin and iron(III) ions. (AO2) `1 mark`

4 a The nickel–cadmium cell produces a voltage of 1.40 V. The cadmium half-cell is the negative electrode. The equations for the half equations are given below.

$$Cd(OH)_2 + 2e^- \rightarrow Cd + 2OH^- \qquad E^\ominus = \text{to be calculated}$$

$$NiO(OH) + H_2O + e^- \rightarrow Ni(OH)_2 + OH^- \quad E^\ominus = +0.52\,V$$

    i Write an overall equation for the reaction which occurs. (AO2) `2 marks`

    ii Calculate the $E^\ominus$ value for the cadmium half-cell. (AO2) `2 marks` iii Write an equation for the reaction which occurs when the cell is recharged. (AO1) 1 mark

b The hydrogen fuel cell operating in alkaline conditions has the overall reaction:

$$2H_2 + O_2 \rightarrow 2H_2O$$

The reaction that occurs at the positive electrode is:

$$O_2 + 2H_2O + 4e^- \rightarrow 4OH^-$$

Write the half equation for the reaction that is occurring at the negative electrode. (AO2)   2 marks

A Brønsted–Lowry acid is a proton donor. A Brønsted–Lowry base is a proton acceptor.

A weak acid is slightly dissociated in solution, but a strong acid is fully dissociated in solution. The slight dissociation of a weak acid is shown using a reversible arrow, for example:

$$CH_3COOH \rightleftharpoons CH_3COO^- + H^+$$

$$pH = -\log_{10}[H^+]$$

[H$^+$] for a strong acid is calculated from its concentration multiplied by its proticity. pH is determined from $-\log_{10}[H^+]$.

K_w is the ionic product of water; $K_w = [H^+][OH^-]$. K_w always has units of mol^2 dm^{-6} and it changes with temperature. K_w at 298 K is 1.00×10^{-14} mol^2 dm^{-6}. The pH of pure water is calculated from its K_w; in pure water [H$^+$] = [OH$^-$], so $K_w = [H^+]^2$, so $[H^+] = \sqrt{K_w}$; pH is calculated using $pH = -\log_{10}[H^+]$.

For a strong alkali, [OH$^-$] is calculated by multiplying the concentration of the alkali by the number of OH$^-$ it produces in solution (which is usually 1). For a strong alkali $[H^+] = K_w/[OH^-]$ where K_w usually is equal to 1.00×10^{-14} unless told otherwise; pH is calculated using $pH = -\log_{10}[H^+]$.

K_a for a weak acid, HA, $= [A^-][H^+]/[HA]$ where [A$^-$] is the concentration of the anion at equilibrium, [H$^+$] is the concentration of hydrogen ions at equilibrium and [HA] is the concentration of the undissociated acid at equilibrium. [H$^+$] for a weak acid is calculated using its K_a; [H$^+$] for a weak acid $= \sqrt{[\text{weak acid}] \times K_a}$. pH is again calculated using $pH = -\log_{10}[H^+]$.

When a strong acid or strong alkali is diluted using water, the new concentration of H$^+$ or OH$^-$ can be calculated from the initial moles and the new final volume. The pH is determined in the usual manner from the new concentration.

When a neutralisation reaction occurs, either H$^+$ or OH$^-$ will be left over. The new concentration of the reactant in excess may be calculated using the moles left over and the new final volume. pH is determined in the usual manner.

A buffer is a solution that resists changes in pH when a small amount of acid or alkali is added. An acidic buffer is a solution containing a weak acid and its salt and is prepared either by mixing directly or by adding a strong base such as NaOH to an excess of a weak acid. A buffer removes added H$^+$ according to the equation:

$$H^+ + A^- \rightarrow HA$$

A buffer removes OH$^-$ according to the equation:

$$HA + OH^- \rightarrow A^- + H_2O$$

The pH of a buffer may be calculated using the K_a expression to determine [H$^+$]; K_a may be given (or pK_a) and [A$^-$] and [HA] calculated from the amounts in the question; pH is determined in the usual manner.

Adding x moles of H$^+$ to a buffer decreases the moles of A$^-$ by x and increases the moles of HA by x, so the new pH of the buffer may be calculated. Adding y moles of OH$^-$ to a buffer decreases the moles of HA by y and increases the moles of A$^-$ by y, so again the new pH of the buffer may be calculated. A basic buffer contains a weak base and its salt, such as ammonia and an ammonium salt.

A titration curve (or pH curve) plots pH against volume of acid or alkali added; its shape is determined by the type of acid and base used (i.e. strong and weak). The vertical region in a titration curve allows us to choose a suitable indicator for a titration.

---

1 Which one of the following is the conjugate acid of HPO$_4^{2-}$? (AO3)　　　　1 mark

A PO$_4^-$

B H$_2$PO$_4^-$

C PO$_4^{3-}$

D H$_2$PO$_4^{3-}$

2 What is the pH of 2.14 mol dm^{-3} sulfuric acid to 2 decimal places? Assume that the sulfuric acid is fully dissociated. (AO2)　　　1 mark

    A　−0.63

    B　−0.33

    C　0.33

    D　0.63

3 Which one of the following always has units of mol dm^{-3}? (AO1)　　　1 mark

    A　K_a

    B　K_c

    C　K_p

    D　K_w

4 Calculate the pH of the following solutions at 298 K. Give all answers to 2 decimal places. (AO2)

$K_w = 1.00 \times 10^{-14}$ mol^2 dm^{-6} at 298 K

    a　0.0125 mol dm^{-3} hydrochloric acid　　　2 marks

    b　0.510 mol dm^{-3} sodium hydroxide solution　　　2 marks

    c　0.248 mol dm^{-3} ethanoic acid ($K_a = 1.80 \times 10^{-5}$ mol dm^{-3})　　　2 marks

5 15.0 cm^3 of a solution of 0.417 mol dm^{-3} sulfuric acid was diluted using deionised water until the pH of the final solution was 1.50. Calculate the volume of water added to the original solution. Give your answer to 3 significant figures. Assume that the sulfuric acid is fully dissociated. (AO2)　　　6 marks

6 The structures below show two weak acids (labelled A and B) together with their pK_a values.

A   B

$pK_a = 2.83$ $pK_a = 3.98$

a i   Name the acids A and B. (AO3)   `2 marks`

A ....

B ....

ii   Explain why acid A exhibits optical activity. (AO1/AO3)   `2 marks`

....

....

....

iii   Write an equation for the reaction of B with sodium carbonate. (AO2)   `1 mark`

....

iv   Suggest why acids A and B are stronger weak acids than propanoic acid. (AO3)   `2 marks`

....

....

....

....

b i   Write an equation for the dissociation of A in aqueous solution. (AO2)   `1 mark`

....

ii   Calculate the pH of a 0.0175 mol dm^{-3} solution of acid A. Give your answer to 2 decimal places. Assume that acid A is not fully dissociated. (AO2)   `3 marks`

....

....

....

....

....

c   25.0 cm³ of the 0.0175 mol dm⁻³ solution of acid A may be titrated using 0.0152 mol dm⁻³ sodium hydroxide solution.

i   Calculate the pH of 0.0152 mol dm⁻³ sodium hydroxide solution at 25°C. Give your answer to 2 decimal places. (K_w = 1.00 × 10⁻¹⁴ mol² dm⁻⁶ at 25°C) (AO2)   3 marks

ii   Calculate the volume of 0.0152 mol dm⁻³ sodium hydroxide solution required for neutralisation. Give your answer to 3 significant figures. (AO2)   3 marks

iii   Which of the titration curves shown below would best represent the titration of acid A using sodium hydroxide solution? (AO3)   1 mark

A

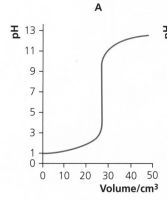

B

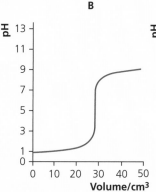

C

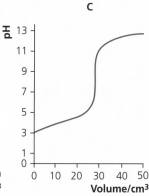

D

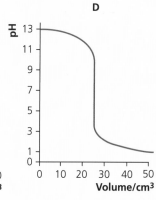

iv   The table below gives information on some indicators.

Indicator	pH range	Lower pH colour	Higher pH colour
Pentamethoxy red	1.2–3.2	Violet	Colourless
Naphthyl red	3.7–5.0	Red	Yellow
4-nitrophenol	5.6–7.0	Colourless	Yellow
Cresol purple	7.6–9.2	Yellow	Purple

Name a suitable indicator for the titration based on the titration curve you have chosen in part iii and state the colour change observed at the end point. Explain your choice of indicator. (AO3)   3 marks

7 Calculate the pH of a buffer solution which contains 0.0125 mol of weak acid HA (K_a = 2.40 × 10^{-4} mol dm^{-3}) and 0.0210 mol of its salt NaA. The total volume of the solution is 100 cm^3. Give your answer to 2 decimal places. (AO2)

3 marks

8 For the buffer solution in question 7, 5.00 cm^3 of 0.104 mol dm^{-3} hydrochloric were added. Calculate the pH of the buffer solution after this addition. Give your answer to 2 decimal places. (AO2)

6 marks

Exam-style question

⏱ 20

9 A solution of 0.242 mol dm^{-3} potassium hydroxide solution is used throughout this question. K_w at 298 K = 1.00 × 10^{-14} mol^2 dm^{-6}. Give all pH answers to 2 decimal places.

a Calculate the pH of a 0.242 mol dm^{-3} potassium hydroxide solution.

2 marks

....

....

....

b 25.0 cm^3 of the potassium hydroxide solution were placed in a conical flask and titrated with 0.140 mol dm^{-3} sulfuric acid.

i Write an equation for the reaction of potassium hydroxide with sulfuric acid.

1 mark

....

....

ii Calculate the volume of 0.140 mol dm^{-3} sulfuric acid required to react with the potassium hydroxide solution. Give your answer to 3 significant figures.

3 marks

....

....

....

....

....

iii Calculate the pH when 15.0 cm^3 of 0.140 mol dm^{-3} sulfuric acid has been added to 25.0 cm^3 of 0.242 mol dm^{-3} potassium hydroxide solution.

6 marks

....

....

....

....

....

....

....

....

....

....

....

....

c 10.0 cm³ of 0.242 mol dm⁻³ potassium hydroxide solution is added to 40.0 cm³ of 0.102 mol dm⁻³ ethanoic acid (K_a = 1.76 × 10⁻⁵ mol dm⁻³) to form an acidic buffer solution. Calculate the pH of the buffer solution formed.

6 marks

Hodder Education, an Hachette UK company, Blenheim Court, George Street, Banbury, Oxfordshire OX16 5BH

Orders

Bookpoint Ltd, 130 Milton Drive, Milton Park, Abingdon, Oxfordshire OX14 4SE

tel: 01235 827827

fax: 01235 400401

e-mail: education@bookpoint.co.uk

Lines are open 9.00 a.m.–5.00 p.m., Monday to Saturday, with a 24-hour message answering service.

You can also order through www.hoddereducation.co.uk

© Alyn G. McFarland and Nora Henry 2016

ISBN 978-1-4718-4505-5

First printed 2016

Impression number 5 4 3 2 1

Year 2020 2019 2018 2017 2016

ISBN 978-1-4718-4505-5